Air Force Cyber Warfare

Now and the Future

Col William J. Poirier, USAF
Maj James Lotspeich, PhD, USAF*

> I think most people today understand that cyber clearly underpins the full spectrum of military operations, including planning, employment, monitoring, and assessment capabilities. I can't think of a single military operation that is not enabled by cyber. Every major military weapon system, command and control system, communications path, intelligence sensor, processing and dissemination functions—they all have critical cyber components.
>
> —Gen William L. Shelton
> Commander, Air Force Space Command

Modern-day cyber warriors are elusive figures. Are they technological ninjas, typing feverishly on a keyboard in a darkened room or perhaps gunslingers throwing cyber bullets

---

*Additional contributors to this article include Col Douglas Coppinger; Lt Col Michael Birdwell, 91 NWS/CC; Lt Col Brian Denman, 690 NSG/CD; Lt Col Paul Williams, 26 NOS/CC; Lt Col Joseph Zell, 33 NWS/CC; Maj Brian Balazs, 26 OSS/DO; Maj Christopher Corbett, 315 NWS/DO; Mr. Richard DeLeon, 26 NOG/TA; and Mr. Richard White, 67 NWW/TA.

downrange at shadowy foes? There are many images of cyber warfare in popular culture. Most of them focus on the individual's uncanny grasp of technology—the ability to exploit any system with a dizzying flurry of keystrokes or to fend off adversaries with a smartphone, a paper clip, and an ingenious plan. These socially awkward heroes and heroines fill the silver screen with visions of a new kind of warfare.

Contradicting these stereotypes, Air Force cyber operations are carefully planned and controlled by disciplined, rigorously trained operators. Rather than acting alone, these professionals produce effects in support of national interests through teamwork, careful coordination, and deliberate, considered targeting based on established national policy. This article discusses the events and thinking that have resulted in today's cyber forces, describes how they operate in cyberspace today, and presents a vision for how they will continue to provide cyberspace dominance in future wars. Although many of the cyber warfare capabilities of tomorrow are speculative in nature, the enabling technologies and policies for them exist today.

A Brief History of Cyber

If we could first know where we are, and whither we are tending, we could then better judge what to do, and how to do it.

—President Abraham Lincoln

Traditionally associated with the explosive growth of network and computing equipment in the 1990s, cyberspace was commonly used to achieve operational objectives during World War II. For example, in the Battle of the Beams, German bombers navigated from continental Europe to Great Britain by following a radio signal transmitted from the point of origin. The pilots would know they were above their targets when they intercepted a second beam, also transmitted from continental Europe. This system ensured that German night raiders found their targets in the dark and returned home safely. British engineers quickly discovered the German use of radio frequency and developed countermeasures. By broadcasting similar signals at precise times, British cyber operators fooled the German bombers, causing them to drop their ordnance at a location chosen by the British. Similarly, the British cyber countermeasures made return trips nearly impossible for the Germans, many bombers never finding home base and a few even landing at Royal Air Force fields, their pilots thinking that they had returned home.[1] This use of the frequency spectrum (a critical portion of cyberspace) to create effects illustrates the operational power of cyberspace long before anyone considered it a domain.[2]

Thus, military operations as far back as World War II incorporated aspects of cyberspace into operations, but almost 60 years passed before leaders formally recognized the importance of this domain. In 2003 President George W. Bush released the *National Strategy to Secure Cyberspace*, followed in 2006 by the *National Military Strategy for Cyberspace Operations*.[3] These two documents established the strategic importance of cyberspace to national interests, but they did not form in a vacuum. To understand how cyberspace began to coalesce conceptually and how leaders began to understand its important role in modern military operations, we must first look at how we've arrived at our current perspective on cyberspace and cyber warfare.

Before cyberspace earned recognition as an operational domain of warfare, the military considered information a target and an instrument of war. In 1993 the Air Force established the Air Force Information Warfare Center (AFIWC) as "an information superiority center of excellence, dedicated to offensive and defensive counter information and information operations."[4] Lessons learned from Operation Desert Storm led to the realization that information is vital to modern military operations and, as such, must be defended from adversaries.[5] By the same token, exploitation of enemy information can be a viable option for gaining an operational advantage.

An attack on Air Force networks by unknown adversaries validated this viewpoint. During the "Rome Lab incident" of March 1994, administrators at Rome Laboratory, New York, found an unauthorized wiretap program—a "sniffer"—on their network that had stolen lab employees' user names and passwords. The attackers—a 16-year-old from the United Kingdom and an unknown person identified only as "Kuji"—successfully obtained information on a number of sensitive defense research projects and used the Rome Lab connection to attack other institutions, stealing all of the data stored on the Korean Atomic Research Institute's computers and depositing it in the Rome Lab computers.[6]

This incident as well other high-profile attacks of the time, such as the theft of data concerning the Strategic Defense Initiative from the Lawrence Berkeley National Laboratory, led to a debate among the Air Force staff regarding whether or not to incorporate the tools and techniques under development at the AFIWC as war-fighter capabilities.[7] On 15 August 1995, the debate ended when the Air Force chief of staff directed development of an information warfare squadron to support Ninth Air Force's combat operations. As a result, the service established the 609th Information Warfare Squadron in October 1995 with a mission to "conceive, develop, and field Information Warfare combat capabilities in support of a Numbered Air Force."[8]

The squadron pioneered defensive counterintelligence operations from 1995 through 1999 and then transferred its mission to the Air Force computer emergency response team, a subdivision of the AFIWC.[9] During this time, a number of events—exercise Eligible Receiver and operations Solar Sunrise and Moonlight Maze—led to an increased interest in information operations at the Department of Defense (DOD) level.[10] Eligible Receiver highlighted critical vulnerabilities in US Pacific Command's systems as well as in 911 and power grids in nine US cities. Analysts were still digesting the results of this exercise when officials discovered attackers stealing tens of thousands of files from systems at the Pentagon, National Aeronautics and Space Administration, and Department of Energy.[11] Detection of additional exploitations of known vulnerabilities in the DOD's unclassified networks further highlighted the need to develop indicators and warnings of attack as well as organize to address weaknesses in information warfare operations.[12]

To address these shortfalls, the DOD activated Joint Task Force–Computer Network Defense under Maj Gen John "Soup" Campbell in December 1998, reporting directly to the secretary of defense and envisioned as having a war-fighting role.[13] In 2000 the task force took on an additional offensive role and a new name—Joint Task Force–Computer Network Operations—to reflect this change. The DOD adjusted the mission again in 2004, this time adding management as well as defense of the department's networks. The offensive mission moved to a new organization, Joint Forces Component Command–Network Warfare.[14] Finally, in 2009 the establishment of United States Cyber Command (USCYBERCOM) rejoined both organizations under a single sub-unified command.[15]

Although the history of cyber is full of organizational changes, we have little documentation of why the military chose to organize as it did to address cyberspace challenges. Attacks on military networks such as Moonlight Maze and Solar Sunrise provide insight only into why defensive operations were necessary, but the organizational changes also reflect a shifting concept of the interactions among defensive, offensive, and network management operations in the realm of cyberspace. Additionally, the evolution from information warfare to cyber warfare indicates a subtle shift in mission: from information as a commodity; to attack and defense of the systems used to process, store, and transmit information; and finally to the domain in which those systems and the information they manipulate reside.

Cyber Warfare Today

Reflecting the military's changing understanding of the nature of cyber warfare, today's operations are defined by a mixture of mature and developing capabilities, doctrine, and organizations. As with air and space domains at their inception, the cyberspace domain continues to mature along a trajectory of increasing capability and capacity; however, many shortfalls exist. Fortunately, military leaders understand them and are sharing their perspective in the national debate. For example, in *Cyber Vision 2025*, Mark Maybury, the former chief scientist of the Air Force, describes the technological, policy, and personnel changes necessary through 2025 to realize future Air Force cyber capabilities.[16] Gen Michael Hayden, USAF, retired, former director of the National Security Agency and Central Intelligence Agency, discusses 10 questions that must be answered before we can truly integrate cyber into national instruments of power.[17] In a recent symposium sponsored by the Armed Forces Communications and Electronics Association, Gen William Shelton, commander of Air Force Space Command, addressed the steps taken by his command to operationalize and integrate cyber forces as well as the issues we face in the near term.[18] Similarly, Maj Gen Suzanne Vautrinot, the former commander of Twenty-Fourth Air Force, now retired, outlined the challenges and strategies for increasing defensive and offensive capabilities in a constrained fiscal environment.[19] The combined efforts of these and many other senior Air Force leaders are driving the maturation of the service's cyber operations by accelerating the pace of innovation.

The Air Force's cyber capability exists on a continuum (see the figure below) ranging from nascent and niche effects to proactive and responsive support of combatant commanders. In today's cyber force, operators occupy the middle of this continuum with niche targets included in operation plans and a mixture of proactive and reactive defensive capabilities. To move combat effectiveness to the right on this chart, the Air Force must implement future initiatives such as US-CYBERCOM's cyber mission force structure and the joint information environment architecture, both of which will enhance the ability of cyber forces to provide theater- and campaign-level support. The Air Force also will continue ongoing initiatives, including Air Force Network (AFNet) migration, and the maturation of cyber weapon systems to increase cyber capacity in terms of the number of missions conducted in support of war fighters.

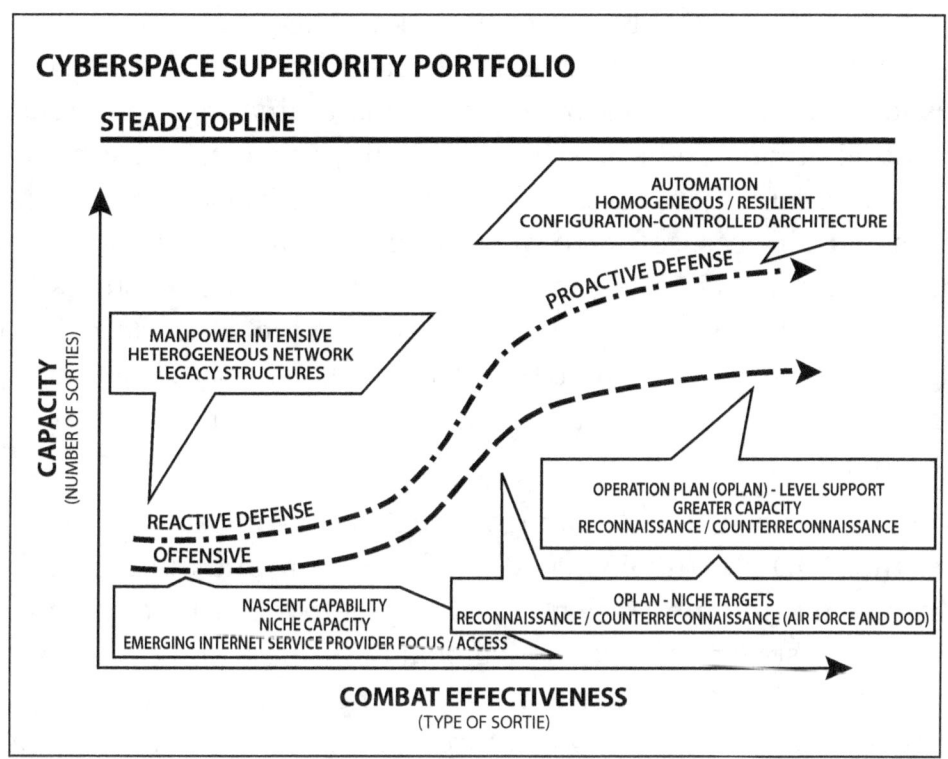

Figure. Cyberspace investment challenge. (Adapted from Maj Gen Suzanne M. Vautrinot, "Sharing the Cyber Journey," *Strategic Studies Quarterly* 6, no. 3 [Fall 2012]: 74, http://www.au.af.mil/au/ssq/2012/fall/fall12.pdf.)

Even though the capability continuum depicts only offensive and defensive cyber forces, modern cyber warfare is conducted by leveraging three operational mission areas: Department of Defense Information Network (DODIN) operations, defensive cyber operations (DCO), and offensive cyber operations (OCO), each of which independently enables effects for the air, space, sea, and land domains.[20] All three are inextricably linked to generate effects across the spectrum of conflict, from small special operations missions to global conventional warfare.

The rapid rise in weapon systems and command and control (C2) systems that rely on network and wireless connections makes the integration and synchronization of complex operations difficult apart from the cyber domain—and underscores the importance to modern military warfare of the DODIN. That network is "the globally interconnected, end-to-end set of information capabilities, and associated processes for collecting, processing, sorting, disseminating, and managing information on-demand to warfighters, policy makers, and support personnel, including owned and leased communication and computer systems and services, software (including applications), data, security services, other associate services, and national security systems."[21] DODIN operations construct, operate, and sustain the cyber domain, offering mission assurance and defense through prioritized network provisioning (dynamic construction), hardening, and configuration management.

Twenty-Fourth Air Force manages the AFNet—the Air Force's portion of the DODIN. With 850,000 total force users and billions of dollars in systems and infrastructure, Twenty-Fourth Air Force's units dynamically construct and operate a global enterprise and provision enterprise services to the Air Force and joint forces worldwide. Additionally, they defend the network through management of both base and AFNet boundaries, sensor placement and management, client configuration, and enterprise-compliance management. The services offered by these units assure that operational planners receive information for missions requiring complex communication topologies, high bandwidth, and high reliability.

Oftentimes people misconstrue DODIN operations as a support or information technology function. For example, Lt Gen Michael Basla, the Air Force's chief information officer, said, "I think we will draw a clearer line and distinction between what is required to build, operate and maintain [Air Force networks] and what is required to operate on the network."[22] Moreover, Gen Mark Welsh, the Air Force chief of staff, has observed that up to 90 percent of Air Force cyber personnel operate Air Force networks and that "they're not what NSA would call a cyber warrior."[23] Although these statements blur the distinction between network maintenance and defense, the DODIN fills an integral role in the conduct of military operations. The obvious benefits include constructing and operating the domain that enables all other domains. Additionally, DODIN operations provision access to information sources, harden friendly portions of the domain from unauthorized access, and configure network systems to provide ease of maneuver to friendly forces while constraining the adversary's options. These actions create a cyber high ground resulting in strategic, operational, and tactical advantages by making mission-critical information easier to defend and harder to attack.

To that point, the Air Force advanced the AFNet's defensive posture through two significant DODIN architecture initiatives. First, the deployment of Air Force gateways reduced the number of external network access points from 120 to 16. This architectural change enabled the service to canalize traffic, characterize the domain, and control data flows to significantly reduce the AFNet attack surface exposed to enemy strikes. The second initiative consolidated 850,000 users into a single integrated Air Force network, enabling enterprise-wide collaboration and improved, trusted secure communications. Defensively, this initiative delivers embedded security that substantially reduces an adversary's ability to act on the network by using compromised user credentials. Collectively, these defensive improvements inverted the cost/risk calculus of attack versus defense by forcing the adversary to work harder to find vulnerabilities while making it easier for the defender to guard critical assets.

The DCO mission area provides active defense against opponents. Twenty-Fourth Air Force's units prevent, detect, and respond to enemy actions through both active and passive defensive capabilities. These units conduct defense through a set of layered, overlapping technologies called "defense in depth," an architecture that ensures monitoring and defense of avenues of access as well as end points such as clients and servers. While DODIN operators limit attack vectors and reduce vulnerabilities by strategic placement of defensive capabilities on the network, DCO operators actively engage adversaries inside Air Force networks to prevent intrusions, detect malicious capabilities and techniques, and respond to system compromises.

DCO operators monitor defenses for signs of attack and configure defenses to foil future attempts. The primary strategy for preventing intrusion calls for detecting known adversary tactics (signatures), limiting visibility into the AFNet, and continuously monitoring intelligence streams for indications of pending attacks. Operators analyze capabilities and methods used by the enemy and develop signatures that match patterns unique to a particular attack and thus provide complete protection from strikes matching the signature. Unfortunately, this method will not block attacks that have been modified from the original salvo. To maneuver around signature-based defenses, cyber attackers must "reengineer" their weapons so that unique signatures compromised in previous attacks are no longer detected. Depending upon the complexity of the developed signature, the adversary may be able to alter his weapons, forcing defenders to develop new signatures. This arms race between attack and defense has traditionally favored the attackers; however, as DODIN forces continue to reduce pathways that opponents can use, and as DCO operators persist in locating and eliminating vulnerabilities, the balance begins to shift in favor of the defense.

When new attacks occur that defenders could not prevent, sensors placed throughout the network supply intrusion indications and point DCO operators to the compromised systems, which they examine (by means of digital forensic analysis) to determine how the intrusion occurred and what tools were used. They then develop countermeasures to prevent future attack. DCO forces remotely access forensic data from all sensor devices to counter future compromises. Defenders use specialized tools to remotely capture the exact state of a computer (e.g., current data in memory, running programs, open network connections, etc.) to determine exactly what is happening at a given moment. This capability takes snapshots of malicious code as it executes, helping defenders understand the exact behavior of implanted software. By analyzing this behavior, they can develop signatures and new tactics, techniques, and procedures (TTP) to prevent the same type of compromise in the future. The use of remote forensics capabilities reduces defenders' incident response from days to hours, slashing the amount of time that attackers have to maneuver through the network, perform reconnaissance, or exfiltrate sensitive data.

Additionally, Twenty-Fourth Air Force has both hunting and pursuit capabilities to offer real-time defense and response against adversary actions and regularly analyze enterprise resources for indications of advanced enemy presence or attempted access. Even though boundary defense is an effective means of recognizing and repelling most attacks, a sufficiently sophisticated and dedicated actor will eventually gain a toehold. Highly skilled DCO operators conduct active pursuit operations to rove the enterprise network and find, fix, track, and target such actors. These operators conduct real-time analysis of network devices, looking for anomalies that indicate enemy activity, eradicating the threat, and initiating an incident-response process to determine the root cause and/or TTPs used to gain access. Sometimes an even more comprehensive look is necessary to ensure that critical assets such as weapon systems and C2 nodes are appropriately hardened and cleared of advanced adversary presence. The Air Force uses hunt operations to characterize the cyber environment in these enclaves, complete a comprehensive analysis of mission data flows, standardize and harden the weapon system or critical asset interfaces, determine potential anomalous activity or attack vectors, herd adversary behavior, and eradicate persistent threats from the environment. These operations, which rely heavily on individual experience, knowledge, and training, are intensive and focused to ensure that these critical assets enjoy freedom of action in contested environments. Even as technology progresses, we will rely heavily on both pursuit and hunting capabilities to counter the advanced adversary threat in the future. Additionally, to increase the capacity and capability of this mission area, USCYBERCOM has developed a cyber protection team structure, each team including a mixture of capabilities designed to give combatant commanders DCO effects. According to Gen Keith Alexander, the commander of USCYBERCOM, the command will stand up 13 teams by the end of 2015, significantly increasing the Air Force's DCO force, strengthening blue networks, and forcing the enemy to divert manpower and attention to counter this new capability.[24]

As with DCO and DODIN operations, OCOs have developed from a nascent to an operational capability well integrated into joint operations. The OCO mission set concentrates on gaining and—more importantly—maintaining access to enemy areas of cyberspace without detection. The nature of OCOs requires operators to carefully plan missions to characterize and exploit enemy networks. Further, the tools used to perform OCOs are sensitive because of the nature of the cyber domain (i.e., the ease of copying bits and bytes). Consequently, tool development and deployment are an important aspect of this mission area.

Although OCO operators provide a very real set of strategic alternatives to combatant commanders, the effects are specific and limited in scope. To exploit an adversary's system, offensive operations demand detailed knowledge of the target network, obtaining such information by performing network reconnaissance with sophisticated TTPs. Once operators have identified vulnerabilities, they must then develop either a technique or a weapon or select one from an existing repository prior to choosing the specific delivery mechanism. After they have accessed their target, operators establish a permanent presence on the machine while cloaking indications of the incursion, allowing them to maintain access indefinitely. Such persistent presence lets them effectively exploit information on the target in support of war fighters' objectives. In light of the long lead time necessary to perform target reconnaissance and establish persistent access, offensive operations typically require advanced planning and a lengthy time horizon to offer effective options.

The weapons used by operators are similar to the ordnance that a pilot employs to carry out a given mission. Certain weapons are better for a desired purpose than others, and some work against a particular set of targets while others are ineffective against that objective. One major difference, however, is their fragility. Since defenders can block a weapon using a signature once they have detected it, use of a given technique or weapon to gain or maintain access carries a risk that the attacker will discover and counter it, rendering the technique or weapon useless for future operations. As a result, operational planners must assess the technical gain/loss associated with the employment of OCOs. If the desired effect is not substantial enough to justify the potential loss of an OCO weapon, then they should consider other methods.

Today's OCO force is a high-demand, low-density asset. As it did with DCOs, to increase the capacity and capability of this mission area, US-CYBERCOM will develop a cyber mission force structure for OCOs, including teams composed of a mixture of capabilities designed to provide a broad spectrum of OCO effects to combatant commanders. General Alexander expects the command to stand up several of these teams by the end of 2015, significantly augmenting the Air Force's OCO force.[25] The increased capacity for OCO operations will put enemy strongholds at risk, forcing adversaries to divert manpower and attention to defenses and reducing the defensive burden on US networks.

The shortfalls of current cyber warfare operations are not readily captured by the dimensions of the capability continuum in the figure depicting the cyberspace investment challenge (see above). Fully illustrating where the cyber domain rests in this continuum requires extending into a third dimension—domain coverage. Contemporary cyber warfare is characterized by largely network-based capabilities in conjunction with traditional electronic warfare. During peacetime, the bulk of the effort focuses on shaping the cyber battlefield, defending critical assets, and collecting intelligence. Should the United States enter a full-scale cyber war today, offensive and defensive capability would be limited to subsets of the full cyberspace domain. These subsets are critical to the projection of power, but they do not fully encompass the overall domain. Such current capabilities, though effective, present limited cyber options to our combatant commanders.

Cyber Warfare in the Future

Victory smiles upon those who anticipate the changes in the character of war, not upon those who wait to adapt themselves after the changes occur.

—Air Marshal Giulio Douhet

Although cyber warfare is currently limited to information networks and network-attached systems, it will drastically expand in the future. Rather than decide between kinetic and nonkinetic effects, planners will choose the effect that will best produce the desired outcome. Cyber-based effects will not be limited to networks of computers; rather, they will encompass all electronic information processing systems across land, air, sea, space, and cyberspace domains. This full-domain dominance will permit freedom of maneuver in all war-fighting domains by holding the enemy's electronic information-processing systems at risk while defending friendly systems from attack.

The future of cyber warfare is predicated on policy, technology, and threat. New technology can have disproportionate effects, not only on the weapons used in cyberspace but also on the makeup of the domain itself. National policy on cyberspace dictates the objectives and rules of engagement for cyber capabilities as well as the organization and execution of operations. The rapidly evolving threat posed by peer actors in the cyber domain will dictate how cyber forces are trained and deployed in the future battlefield. Despite these wildcard influences, the future of cyber warfare can be broadly extrapolated from current experience and application of fundamental tenets of warfare. To remain grounded in today's realities, we limit the vision of cyber warfare discussed here to a decade into the future, allowing us to assume that technological changes will follow the course laid out in *Cyber Vision 2025*.[26]

Future cyber warfare will not be relegated solely to network-based resources. According to Major General Vautrinot, "Cyberspace is not simply the Internet; rather, it is a network of interdependent information technologies including the Internet, telecommunications networks, computer systems, and embedded processors."[27] Although much of the present effort focuses on Internet-connected networks, this is only a subset of the total cyber domain, which also includes non-Internet-connected networks such as tactical data links, satellite-control networks, launch-control networks, and other networks not traditionally based on Internet data-transfer protocols and technologies. Future warfare will see DODIN operations as well as DCO and OCO forces expanding their mission areas to these nontraditional networks and the systems that connect through them, such as satellites, avionics, targeting pods, digital radios, and remotely piloted aircraft. Effects produced on and through these systems will include disruption, distraction, distortion, distrust, confusion, and chaos of both a virtual and physical nature, with consequences that can be assessed and measured on the battlefield.

In this future war, many of the services currently supplied by DODIN operations will be decoupled from the hardening, defense, and mission-assurance roles. Services such as e-mail, data storage, web, and transport will be provided as commodity services/utilities, much like electricity or water. Through the joint information environment, the DOD will leverage economies of scale and cloud technologies to improve the resiliency of services and expand their reach so the war fighter can safely assume availability and reliability. This roll-up of commodity services will free DODIN operators to concentrate on defensive hardening and attack recovery while expanding their scope to nontraditional networks. As with AFNet, consolidation and standardization of tactical and C2 networks will result in a reduced attack surface, higher reliability, and more responsive disaster recovery. Rather than rely on weapon system designers to take responsibility for the security of their systems, DOD professionals will manage and enforce formalized security standards and interoperable interfaces. The standards will ensure that weapon systems have a "baked-in security" capability while the interoperable interfaces will reduce the "one-off" systems and capabilities that drive increased enterprise vulnerabilities and cost. Sensors, reporting mechanisms, and configuration-management tools will be designed into the system from the beginning, allowing DODIN operators to enforce a rigorous and standard security posture across all combat systems.

Future DCO capabilities will tackle one of the greatest costs associated with defense: the man-in-the-loop sensor, which offers alerts that require human intuition and experience to interpret and identify the occurrence of a compromise. This reliance on human intuition forces defenders to maintain large, well-trained manpower pools to defend relatively small areas of cyber terrain. The human limitation prevents analysis of these alerts at the speed of data passing through the network, forcing defenders to react to threats rather than proactively defeat them. As technology advances, the infusion of human intuition into automated sensors will allow for man-on-the-loop defense, which will reduce manpower requirements but increase overall effectiveness.

Building upon a standardized security framework, future DCO capabilities and sensors—deployed across all combat platforms—will be designed to supply man-on-the-loop rather than man-in-the-loop detection. These sensors will leverage machine-learning techniques and predictive-behavior modeling to recognize and separate attacks from normal operational data flows. Rather than rely on a human to view and interpret results, defenders will mitigate attacks on the fly and ignore false positives, with human intervention driven by triggers and confidence thresholds.[28] Using ubiquitous network sensors, they will also perform data correlation and analysis across platforms and networks to discover trends of attacks, using them to further characterize current and emerging adversary tactics and give some perspective on both persistent and fleeting targets of enemy interest.

Armed with information on targets under attack in cyberspace, defenders will perform critical asset protection. Expansion outside traditional networks will require that defenders focus on prioritized assets, a process enabled by situational awareness tools that tie missions to systems and physical locations to network locations. Defenders need not protect every workstation equally; instead, they can focus their efforts on systems supporting a high-priority operation or on data links critical to attaining a war fighter's objective. This prioritization of effort will allow them to utilize both mass and maneuver to best counter enemy actions in a timely and effective manner.

Improved sensors and prioritized defenses will allow defenders to push enemy actors outside blue cyberspace. Today's defense in depth catches many attacks inside the boundaries of our networks. In the future, improved sensor capabilities, combined with automated responses, will frustrate most attacks at the boundary of blue space, letting defenders focus on identifying threats before they reach friendly cyber systems and reporting the threats to offensive forces early enough for OCO operators to conduct operations, if necessary. By increasing the engagement distance, defenders will ensure system and data integrity and force attackers to battle through offensive interception before they can attempt to attack friendly systems.

Building on the capabilities of DCOs, future OCO capabilities will split into two types of missions: interception and attack. The former will engage enemy actors as they prepare to strike friendly forces whereas attack missions will hold enemy assets at risk in their own areas of cyberspace. Each mission will engage enemies on both traditional and nontraditional networks in the cyber domain.

Interceptor missions act in conjunction with DCO sensor targeting to attack enemies before they reach friendly systems. These missions will harass the enemy by capturing tools before he can launch them, changing attack targets so that his tools attack the wrong system or commit fratricide, and manipulating the data presented to the enemy operator, forcing him to react to forged threats. Rapid forensic capabilities let defenders reverse-engineer tools captured by interceptors and apply defenses against those tools in real time, foiling any further attempts. These interceptor missions will represent a close air support function in cyber that keeps friendly cyberspace safe by attacking the threat before it arrives.

Attack missions, on the other hand, represent the strategic strike capability of OCOs and will create both virtual and physical effects across all domains through application of offensive capabilities in the cyber domain. Virtual effects will include manipulating data on enemy C2, intelligence, surveillance, and reconnaissance systems; injecting false data into C2 networks and tactical data links; removing data from those links; and isolating systems from their associated networks. Physical effects might include destruction through manipulation of digital control systems or remote system control of platforms such as satellites, remotely piloted aircraft, and fly-by-wire systems. In addition to these effects, attack will provide intelligence collection, data exfiltration, and other more traditional capabilities, but these will be employed across the cyber domain to include satellite systems, aircraft, and C2 systems.

In support of the full-domain competencies discussed above, cyber operators will have comprehensive situational awareness of the cyber domain. Although traditional sensors permit monitoring of the avenues of ingress and egress and small subsets of endpoint behavior, it will be necessary to develop new sensors that alert defenders to behavioral anomalies or statistically significant departures from the expected baseline. Sensors will supply these alerts in an actionable form so that operators can quickly determine whether or not a large-scale attack is occurring or a single node is compromised. Additionally, it will be possible to visualize the cyber domain in terms of logical connections, such as network and radio frequency circuits supporting a given mission, or data flows supporting a desired mission area to provide mission assurance.

Current cyber sensors utilize priorities associated with specific alerts to warn operators of possible malicious action. To determine whether or not those alerts represent a true threat or merely a false positive,

DCO operators must review detailed information such as the actual data passing between computers, the machines involved in the suspect transaction, and the basis of the original alert. This time-intensive process requires highly skilled operators and is prone to human error. Additionally, the alerts signify singular events that occur in a stream of data and may occur ambiguously under normal operating conditions as well as during an attack.

Future situational awareness tools, though, will capitalize on advanced threat indicators such as divergence from expected behaviors. These sensors will use a known baseline of user activity on a given node to determine whether or not a node is deviating from its expected behavior. Using a defense-in-depth methodology, sensors will automatically correlate similar behavioral alerts across multiple clients. With this type of automation, DCO operators can validate alerts at a higher level, in less time, and with reduced manpower. Moreover, behavioral alerting will decrease the number of false positives produced by sensors, allowing operators to spend more time responding to real incidents rather than analyzing nonevents.

Operators will receive alerts in an actionable form. For example, if a sensor alerts them to possible data exfiltration, it will automatically store the data stream in a temporary buffer pending operator action. If the operator confirms the alert, then the act of confirmation will delete the data in question before it is delivered; if the operator determines that the alert is a false positive, then the transmission will be resumed with no data loss. Similarly, attempts to compromise an aircraft or a satellite data link will result in an operator alert indicating the source of the attempt, methods used, and possible attribution based on known TTPs. This level of situational awareness enables the operator to alert the component commander in a timely manner so that he or she can take appropriate kinetic or nonkinetic action in response to the attack.

Finally, situational awareness tools will offer both physical and logical mapping of data and nodes. Since the cyber domain contains both data and the nodes that process it, many parts of the domain possess both a physical and a logical location. For example, systems used to perform space launch may reside at an Air Force base thousands of miles from the actual launch location. A cyber situational awareness tool must be able to depict the systems both as a physical device associated with a given location and as a logical portion of the space launch network. It is also necessary to visualize data flow so that operators can see where spikes in data flow occur, where data is diverted for unknown reasons, and where it has stopped flowing. The increased visualization of data traversing cyberspace will permit operators to better understand and react to changes in both the physical and virtual battlespace.

To conduct cyber operations across the entire domain, we will develop Airmen with the foundational knowledge to comprehend traditional Internet-protocol-based networks as well as radio-frequency and proprietary-communications networks. Further, these warriors must understand not only how devices that operate in the cyber domain are designed but also how they operate. Just as a pilot must have knowledge of aerodynamic fundamentals to understand the performance and limitations of his weapon system, so must cyber warriors possess a foundational grasp of the cyber domain to employ cyber weapon systems properly.

As in the air and space domains, successful deployment of weapon systems in a combat environment demands that cyber crews develop competency in these weapons over the course of a career. Doing so requires a career-field-management strategy that emphasizes the development of experience and expertise tied to weapon system employment. Much like pilots, cyber warriors will be assigned to a mission track (e.g., DODIN operations, DCOs, or OCOs) and a weapon system. During initial qualification training, operators will become proficient in the configuration, components, design, and operation of their system. Over the course of one or more operation tours, they will continue to build expertise and competence in the deployment of that weapon system. Like members of the flying community, those operators will have opportunities to transition to different systems as well as serve on staff or career-broadening tours. Each career path will remain generally distinct in technical development yet emphasize leadership, supervision, and cooperative action that translates to broader Air Force and joint operational expertise over time. The necessary skills and experience will be normalized with the joint community to ensure that forces presented to combatant commanders provide reliable capabilities consistent with those of the other services.

The Air Force will train cyber operators in a rigorous, deliberate fashion to ensure that they possess the foundational skills to perform their specific mission. This training will encompass networking and computing fundamentals as well as knowledge of data transmission across the electromagnetic spectrum, operating systems, computer design fundamentals, and electronic circuit theory. Training specific to mission areas will encompass not only particular toolsets but also defensive and offensive techniques. Both DCO and OCO personnel will routinely rotate into DODIN positions to guarantee current knowledge of system configuration, defensive posture, and terrain familiarization.

Conclusion

Just as the air and space domains took time to grow from their inceptions to fully capable war-fighting domains, so is the cyber domain poised to follow the same arc. That domain has developed at a rapid pace from a novelty and mission-enhanced commodity to a mission-critical capability in just a few decades. As it continues to progress, the level of capability offered by dedicated operators to the war fighter will also increase exponentially.

We can compare today's cyber power to airpower sometime during the interwar years. Operators have developed capabilities and demonstrated their effectiveness to combatant commanders; however, warfare in and through cyberspace remains underdeveloped. Even though professionals in the cyber field have become more proficient at creating effects in the domain via DODIN as well as DCO and OCO operations, these effects are still not well integrated into a combat environment. As was the case with airpower before the beginning of World War II, operational planners are not sufficiently versed in this domain to intuitively envision cyber's contribution to decisive battlefield effects in modern form. Partly because of occasional doubt regarding the proficiency of cyber capabilities, their effects are currently considered "nonkinetic" while more traditional military capabilities produce "kinetic" effects. In the future, cyber warfare will prove its effectiveness on par with more traditional capabilities, blurring the line between kinetic/nonkinetic effects. By then, cyber capabilities will have become well-deliberated strategic alternatives for our national leaders and combatant commanders—recall World War II's Battle of the Beam, mentioned above, when cyber capabilities were the first and best option to defend Great Britain against German bombing raids.

The explosive growth in cyber today and the bold vision articulated by senior leaders throughout the DOD promise a bright future for this domain. As cyber warriors continue to develop competence and effectiveness in their weapon systems, the capabilities they bring to the joint fight will begin to show their true potential. As we plan and employ such capabilities with greater frequency and effectiveness, commanders will fully understand how best to utilize these forces to fulfill mission objectives. Advances in technology, organization, and operator expertise will continue to translate into unprecedented battlefield effects. ✪

Notes

1. Alfred Price, *Instruments of Darkness: The History of Electronic Warfare* (London: Panther, 1979), 55–58.
2. Air Force Doctrine Document (AFDD) 3-12, *Cyberspace Operations*, 15 July 2010, 2–3, http://static.e-publishing.af.mil/production/1/af_cv/publication/afdd3-12/afdd3-12.pdf.

3. Department of Homeland Security, *The National Strategy to Secure Cyberspace* (Washington, DC: Department of Homeland Security, February 2003), http://www.defense.gov/home/features/2010/0410_cybersec/docs/cyberspace_strategy[1].pdf; and Chairman of the Joint Chiefs of Staff, *The National Military Strategy for Cyberspace Operations* (U) (Washington, DC: Chairman of the Joint Chiefs of Staff, December 2006), http://www.dod.mil/pubs/foi/joint_staff/jointStaff_jointOperations/07-F-2105doc1.pdf.

4. Joseph A. Ruffini, "609 IWS Chronological History," 609 Air Operations Squadron, June 1999, 8; and "Air Force Information Warfare Center," in *Air Intelligence Agency Almanac*, Federation of American Scientists, accessed 21 June 2013, http://www.fas.org/irp/agency/aia/cyberspokesman/97aug/afiwc.htm.

5. Armed Forces Communications and Electronics Association, *The Evolution of U.S. Cyberpower* (Fairfax, VA: Armed Forces Communications and Electronics Association, n.d.), 11–17, http://www.afcea.org/committees/cyber/documents/TheEvolutionofUSCyberpower.pdf.

6. Senate, *Committee on Governmental Affairs, Permanent Subcommittee on Investigations (Minority Staff Statement), Hearing on "Security in Cyberspace,"* 104th Cong., 2nd sess., 5 June 1996, app. B.

7. Cliff Stoll, *The Cuckoo's Egg: Tracking a Spy through the Maze of Computer Espionage* (New York: Doubleday, 1989), 180–86; and Ruffini, "609 IWS Chronological History," 3.

8. Ruffini, "609 IWS Chronological History," 4.

9. Ibid., 4–30.

10. "Cyberwar!," *Frontline* (Public Broadcasting Service), 24 April 2003, http://www.pbs.org/wgbh/pages/frontline/shows/cyberwar/; and "Solar Sunrise," GlobalSecurity.org, accessed 21 June 2013, http://www.globalsecurity.org/military/ops/solar-sunrise.htm.

11. "Cyberwar!"

12. "Solar Sunrise."

13. Jason Healy, "Claiming the Lost Cyber Heritage," *Strategic Studies Quarterly* 6, no. 3 (Fall 2012): 12–13, http://www.au.af.mil/au/ssq/2012/fall/fall12.pdf; and History, Defense Information Systems Agency, Department of Defense, 2000s, accessed 5 August 2013, http://www.disa.mil/About/Our-History/2000s.

14. History, Defense Information Systems Agency.

15. Jason Healy and Karl Grindal, "Lessons from the First Cyber Commanders," *New Atlanticist*, Atlantic Council, March 2012, http://www.acus.org/trackback/65665.

16. Mark T. Maybury, PhD, *Cyber Vision 2025: United States Air Force Cyberspace Science and Technology Vision, 2012–2025*, AF/ST TR 12-01 (Washington, DC: Office of the Chief Scientist, United States Air Force, 2012), http://www.af.mil/shared/media/document/AFD-130327-306.pdf.

17. Gen Michael V. Hayden, "The Future of Things 'Cyber,' " *Strategic Studies Quarterly* 5, no. 1 (Spring 2011): 3–7, http://www.au.af.mil/au/ssq/2011/spring/spring11.pdf.

18. Gen William L. Shelton, commander, Air Force Space Command (remarks, Armed Forces Communications and Electronics Association Cyberspace Symposium, Colorado Springs, CO, 6 February 2013), http://www.afspc.af.mil/library/speeches/speech.asp?id=728.

19. Maj Gen Suzanne M. Vautrinot, "Sharing the Cyber Journey," *Strategic Studies Quarterly* 6, no. 3 (Fall 2012): 71–87, http://www.au.af.mil/au/ssq/2012/fall/fall12.pdf.

20. Cheryl Pellerin, "Cyber Command Adapts to Understand Cyber Battlespace," Department of Defense, 7 March 2013, http://www.defense.gov/news/newsarticle.aspx?id=119470.

21. Joint Publication 3-12, *Cyberspace Operations*, 5 February 2012, 18.

22. John Reed, "What Does Cyber Even Mean?," *Killer Apps* (blog), 5 December 2012, http://killerapps.foreignpolicy.com/posts/2012/12/05/what_does_cyber_even_mean.

23. Ibid.

24. Ellen Nakashima, "Pentagon Creating Teams to Launch Cyberattacks As Threat Grows," *Washington Post*, 12 March 2013, http://articles.washingtonpost.com/2013-03-12/world/37645469_1_new-teams-national-security-threat-attacks.

25. Aliya Sternstein, "Military Cyber Strike Teams Will Soon Guard Private Networks," Nextgov, 21 March 2013, http://www.nextgov.com/cybersecurity/cybersecurity-report/2013/03/military-cyber-strike-teams-will-soon-guard-private-networks/62010/.

26. Maybury, *Cyber Vision 2025*.

27. Vautrinot, "Sharing the Cyber Journey," 75.

28. Recent advances in support vector machines specifically and machine learning / classification tasks in general support this assertion. See "Unsupervised Learning and Clustering," in Richard O. Duda, Peter E. Hart, and David G. Stork, *Pattern Classification and Scene Analysis: Part I, Pattern Classification*, 2nd ed. (New York: John Wiley & Sons, 1995).

Col William J. Poirier, USAF

Colonel Poirier (BS, University of Massachusetts–Amherst; MS, Strayer University; MS, National War College) commands the Air Force's newest combat wing, the 67th Network Warfare Wing, Joint Base San Antonio–Lackland AFB, Texas. His wing presents trained and ready cyber forces through Air Forces Cyber to US Cyber Command and other joint task force and combatant commanders to execute global network operations, defense, and full-spectrum network warfare capabilities. The 67th Network Warfare Wing also performs electronic systems security assessments to improve operational security for the Air Force and joint partners. During his career, he has commanded a squadron and has completed assignments at the Air Staff, two unified combatant commands, a binational command, and a defense agency. He has also led two divisions on the staff of a combined force air component commander; served as the chief command, control, communications, computers, and intelligence (C4I) engineer in Operations Southern Watch, Enduring Freedom, and Iraqi Freedom; enabled launch and maintenance operations for strategic and tactical nuclear weapon systems; and was a member of the start-up team for United States Northern Command. Colonel Poirier is a graduate of the Army Command and General Staff College and National War College.

Maj James Lotspeich, PhD, USAF

Major Lotspeich (USAFA; MS, Air Force Institute of Technology; PhD, Naval Postgraduate School) serves as director of operations for the 33d Network Warfare Squadron. He trains and readies cyber forces for presentation through Air Forces Cyber to US Cyber Command and other joint task force and combatant commanders to execute defense of the Air Force portion of the Department of Defense's global enterprise network. He is responsible for directing all Air Force network defense operations in support of the Air Force Network Operations commander and US Cyber Command. Major Lotspeich has held a variety of leadership positions at the base and major command levels, both in-garrison and deployed, serving as leader of postal operations for Air Combat Command and as mission systems flight commander during Operation Iraqi Freedom. Additionally, Major Lotspeich served as assistant professor of computer science at the US Air Force Academy where he directed the core computer science course for 1,400 cadets each year.

Let us know what you think! Leave a comment!

Distribution A: Approved for public release; distribution unlimited.

Disclaimer

The views and opinions expressed or implied in the *Journal* are those of the authors and should not be construed as carrying the official sanction of the Department of Defense, Air Force, Air Education and Training Command, Air University, or other agencies or departments of the US government.

This article may be reproduced in whole or in part without permission. If it is reproduced, the *Air and Space Power Journal* requests a courtesy line.

http://www.airpower.au.af.mil

www.ingramcontent.com/pod-product-compliance
Lightning Source LLC
Chambersburg PA
CBHW081823170526
45167CB00008B/3520